You Are Probability

Surfing the Matrix

Also available from the MSAC Philosophy Group

Spooky Physics

Darwin's DNA

The Magic of Consciousness

The Gnostic Mystery

When Scholars Study the Sacred

Mystics of India

The Unknowing Sage

String Theory

In Search of the Perfect Coke

Is the Universe an App?

You are Probability

Surfing the Matrix

Mt. San Antonio College
Walnut, California

First Printing: 2014

ISBN: 978-1-56543-804-0

MSAC Philosophy Group
Mt. San Antonio College
1100 Walnut, California 91789 USA

Website: http://www.neuralsurfer.com

Imprint: *The Runnebohm Library Series*

Dedication

To our two boys:

Shaun-Michael and Kelly-Joseph

Table of Contents

Acknowledgements

Andrea and I would like to express our deepest thanks to Frank Visser and *Integral World* for publishing a large number of our articles over the years and for encouraging us in exploring the frontiers of neuroscience, quantum physics, and evolutionary biology.

The MSAC Philosophy Group

MSAC Philosophy Group was founded at Mt. San Antonio College in Walnut, California in 1990. It was designed to present a variety of materials--from original books to essays to websites to forums to blogs to social networks to films--on science, religion, and philosophy. In 2008 with the advent of print on demand and cloud computing, the MSAC Philosophy Group decided to embark on an ambitious program of publishing a large series of books and magazines. Today there are well over 100 distinct magazine titles and 50 book titles. In addition, the entire MSAC database is now being put online via Amazon's Kindle, Barnes and Noble's Nook, Google's eBooks, and Apple's iBooks. A special mobile app called Neural Surfer Films is now available for Apple's iPhones and iPads, as well as one for Android operating systems on smart phones and tablets. *The Runnebohm Library* contains works on Einstein, Turing, Russell, Crick and other luminous thinkers. Some of the more popular titles include, *Darwin's DNA: A Brief Introduction to Evolutionary Philosophy* and *Global Positioning Intelligence: The Future of Digital Information*. Finally, *The Runnebohm Library* is in the process of producing a number of highly interactive texts that will include embedded video, games, and interactive feedback loops.

1 | *Desultory Decussation*

I enjoyed reading Eliot Benjamin's intriguing essay, *License Plate Synchronicity*, because he graphically demonstrates how apparently random outer events can coincide with our own subjective wants and needs and produce tremendously powerful intersections.

Eliot Benjamin in his conclusion suggests that such synchronicities "may serve as a reminder that there is indeed inherent spirituality in the universe that cannot be explained by our rational scientific technological minds or brains."

While I understand that unusual occurrences can indeed be interpreted in super-mundane ways, it doesn't mean that such events are the result of something trans-rational. Indeed, even the most apparently miraculous of synchronicities may have a mathematical basis.

The probability of any two events intersecting in meaningful ways is higher than we usually suspect. The linchpin in all of this is our ability to remain aware of how probabilities arise in our life, moment-to-moment, hour-to-hour, and day-to-day.

John Edensor Littlewood, one of the great mathematicians of the last century and a senior wrangler at Cambridge University worked intensively on the theory of large numbers. In his extensive research, oftentimes partnering with his more famous cohort G.H. Hardy, Littlewood unearthed some remarkable properties in large numbers that at first glance seem extraordinarily odd. One peculiar oddity is what is now known as *Littlewood's Law of Miracles*.* Freeman Dyson writing in the *New York Review of Books* explains it this way: "Littlewood's Law of Miracles states that in the course of any normal person's life, miracles happen at a rate of roughly one per month. The proof of the law is simple. During the time that we are awake and actively engaged in living our lives, roughly for eight hours each day, we see and hear things happening at a rate of about one per second. So the total number of events that happen to us is about thirty thousand per day, or about a million per month. With few exceptions, these events are not miracles because they are insignificant. The chance of a miracle is about one per million events. Therefore we should expect about one miracle to happen, on the average, every month. Broch tells stories of some amazing coincidences that happened to him and his friends, all of them easily explained as consequences of Littlewood's Law."**

Or, as framed in an Internet posting, one commentator wrote, "Succinctly put, the law of truly large numbers states: With a large enough sample, any outrageous thing is likely to happen. The point is that truly rare events, say events that occur only once in a million [as the mathematician Littlewoood (1953) required for an event to be surprising] are bound to be plentiful in a population of 250 million people. If a coincidence occurs to one person in a million each day, then we expect 250 occurrences a day and close to 100000 such occurrences a year. Going from year to a lifetime and from the population of the United States to that of the world (5 billion at this writing), we can be absolutely sure that we will see incredibly remarkable events. When such events occur, they are often noted and recorded. If they happen to us or someone we know, it is hard to escape that spooky feeling."

There is another branch off of Littlewood's theory of large numbers that underpins Eliot Benjamin's experiences, which I call *Desultory Decussation* (where two apparently random events intersect to form an X).

If there are thousands, nay millions, of events in our lives (measured in transparently fractal ways), then it should be expected that for every 10,000 plus events, there might be two or more events that intersect. Notice that intersection and you will be aware of a meaningful coincidence—the meaning being that two disparate parts have something in common (whatever that intersection may entail).

We can even splinter off from this and make a broad sweeping generalization. There are those who look or seek out these desultory decussations and those who do not. I would imagine that some of us are more attuned or keenly aware of the intersections (which happen randomly) and they will end up seeing more meaning in their lives, even if the meaning quota is the same relatively speaking for all.

2

In other words, there are those who seek the Littlewood stream and plunge right in and those who do not. Blind typing may in fact produce a legible word just by chance, but the key in all this is to actually become aware of that probability and notice it when such does occur. Otherwise, so many amazing happenings of chance go by completely undetected.

If we could remain conscious of this mathematical matrix, we could be experiencing stunning hierophanies not only monthly, but perhaps daily. We already know that the theory of large numbers bears this possibility out. The only real glitch resides within our selves. To experience Littlewood miracles or desultory decussations (random events interwining in meaningful X patterns), it takes a Herculean effort on our part to remain open to what strange coincidences nature may throw out at us. Littlewood's Law, interestingly enough, first requires us to be attentive, exceptionally so. I think Eliot Benjamin's experiment with license plates illustrates this quite nicely.

I would suggest that a modified version of Littlewood's Law, similar to what we define as desultory decussation, could explain anomalous synchronicities and therefore we do not need to invoke spiritual or paranormal theories for them. *The high improbability of an event oftentimes blinds us from the probability, even if rare, that such events are probabilistic.*

NOTES

* The law was framed by Cambridge University Professor J. E. Littlewood, and published in a collection of his work, *A Mathematician's Miscellany*; it seeks (among other things) to debunk one element of supposed supernatural phenomenology and is related to the more general Law of Truly Large Numbers, which states that with a sample size large enough, any outrageous thing is likely to happen. Littlewood defines a miracle as an exceptional event of special significance occurring at a frequency of one in a million. He assumes that during the hours in which a human is awake and alert, a human will experience one event per second, which may either be exceptional or unexceptional (for instance, seeing the computer

screen, the keyboard, the mouse, the article, etc.). Additionally, Littlewood supposes that a human is alert for about eight hours per day. As a result, a human will, in 35 days, have experienced, under these suppositions, 1,008,000 events. Accepting this definition of a miracle, one can be expected to observe one miraculous occurrence within the passing of every 35 consecutive days – and therefore, according to this reasoning, seemingly miraculous events are actually commonplace. *--Wikipedia entry on Littlewood's Law*

** Dyson, Freeman. "One in a Million." *In The Scientist as Rebel. New York Review of Books*: New York, 2006; paperback ed., pub. 2008: p. 327

2 | Apophenia and the Intentionality Fallacy

"The theory of probabilities is at bottom nothing but common sense reduced to calculus; it enables us to appreciate with exactness that which accurate minds feel with a sort of instinct for which of times they are unable to account."
--Pierre Simon Laplace

I appreciate Elliot Benjamin's recent attempt to justify his belief in mysterious synchronicities and his elaborations on why his personal experiences do not seem to be the result of Littlewood's Law of Miracles or even *Desultory Decussation*.

I agree with him, in part, because a close analysis of his license plate encounters clearly points to an easier, even more rudimentary, explanation for the phenomenon. Contrary to what Elliot Benjamin may wish to argue, the details he presents shows quite clearly that the major factor determining these so-called "spiritual" intersections is his own desire to find meaning in seemingly random events. Quite frankly, Elliot is projecting and transferring his own intentions upon a series of license plates and then deriving some purpose or design as to why those letters or numbers have some special significance.

Human beings do this all the time since it part and parcel of what it is to be human. We are meaning seekers creatures and we are predisposed to find meaningful patterns in all sorts of events, even if those events are random in nature.

As Michael Shermer, founder of *Skeptic Magazine* and the author of *Why People Believe in Weird Things*, writes: "Humans are pattern-seeking, storytelling animals. We look for and find patterns in our world and in our lives, then weave narratives around those patterns to bring them to life and give them meaning. Such is the stuff of which myth, religion, history, and science are made. Sometimes the patterns we find represent reality — DNA as the basis of heredity or the fossil record as the history of life. But sometimes the patterns are imposed by our minds rather than discovered by them — the face on Mars (actually an eroded mountain) or the Virgin Mary's image on the side of a glass building in Clearwater, Florida (really an oil stain from a palm tree, since removed to enable the faithful to better view their icon). The rub lies in distinguishing which patterns are true and which are false, and the essential tension (as Thomas Kuhn called it) pits skepticism against credulity as we try to decide which patterns should be rejected and which should be embraced."

A close analysis of the first four license plate synchronicities that Elliot Benjamin provides tells us more about his own predilections than it does about some transmundane occurrence. Ironically, the mathematics that he invokes to objectively substantiate the improbability of his encounters are, in fact, a complete ruse. Probabilities and the like have actually nothing to do with the synchronicities, since the real issue at hand is Elliot's own pattern seeking. What is at work here is the intentionality fallacy, where we as subjective creatures conflate our subjective needs and wants with outward events, mistakenly believing that the latter is literally contouring to our internal forms of awareness. Elliot Benjamin hasn't tapped into some integral networking of the divine that supersedes rational science. No, what he has uncovered is how easy it is to confuse one's neurology for ontology and then pass it off as beyond current scientific explanation.

What is most important in Elliot's examples is what he leaves out, since it is in those middling details that we can uncover how intentionality, and not some heavenly agent, is at play. Richard Feynman, the well-known architect of Quantum Electrodynamics, strenuously argued that a scientist should point out those details which may contradict his findings or allow others to reevaluate his results in new and perhaps contrarian ways. As Feynman so persuasively argued in his now famous "Cargo Cult" Speech at Cal Tech in 1974,

"It is interesting, therefore, to bring it out now and speak of it explicitly. It's a kind of scientific integrity, a principle of scientific thought that corresponds to a kind of utter honesty--a kind of leaning over backwards. For example, if you're doing an experiment, you should report everything that you think might make it invalid--not only what you think is right about it:

other causes that could possibly explain your results; and things you thought of that you've eliminated by some other experiment, and how they worked--to make sure the other fellow can tell they have been eliminated. Details that could throw doubt on your interpretation must be given, if you know them. You must do the best you can--if you know anything at all wrong, or possibly wrong--to explain it. If you make a theory, for example, and advertise it, or put it out, then you must also put down all the facts that disagree with it, as well as those that agree with it. There is also a more subtle problem. When you have put a lot of ideas together to make an elaborate theory, you want to make sure, when explaining what it fits, that those things it fits are not just the things that gave you the idea for the theory; but that the finished theory makes something else come out right, in addition. In summary, the idea is to try to give all of the information to help others to judge the value of your contribution; not just the information that leads to judgment in one particular direction or another."

THE INTENTIONALITY GAME

Elliot Benjamin provides the following as his first example and reminder of what he claims "cannot be explained by our rational scientific technological minds or brains."

Writes Benjamin, "A few days ago, while I was agonizing over having recently lost one of my mental health jobs, I found myself driving behind a license plate that said ACT. For me this was an immediate recognition of the meaningful workshop I had done a few years ago in Acceptance and Commitment Therapy (Hayes, Strousahl, & Wilson, 2004), which is abbreviated as ACT. In ACT you are taught to accept your disappointments and difficulties in life in a mindful way, and then make a commitment to actualizing your deepest values in life in spite of these disappointments and difficulties (Hayes, Strosahl, & Wilson, 2004). Seeing the ACT license plate was a meaningful reinforcement for me that I needed to accept the loss of my mental health job gracefully, and was connected to my deepest intention of offering my services to continue to

7

work with mental health clients, independently and without expecting to earn any real money from doing so."

Let's systematically breakdown Elliot's example here which I think will easily illustrate how and why intentionality (and not some mysterious spiritual interplay) is at work. First, Elliot reveals his emotional condition with these words "I was agonizing over having recently lost one of my mental health jobs." Thus, this is the mental context in which to better understand what he reports next, "I found myself driving behind a license plate that said ACT."

Importantly, the interested reader will not notice anything of significance or that what just transpired constituted a synchronicity. Rather, it is Elliot who draws a meaning out of these three letters, which he then conjures up as reminding him of a workshop he done in "Acceptance and Commitment Therapy."

At this juncture, several problems arise (especially in light of Feynman's insistence about revealing more, versus less, information which could cast one's own pet theory in doubt), not the least of which is that those three letters ACT could be interpreted in a variety of ways. The fact that Elliot finds meaning in them is precisely the point: it is his own projection from his mental state onto those letters. The letters themselves don't indicate their meaning objectively in a way that others may agree with Elliot at all. That Elliot may find significance in those three letters is one thing, but to then extrapolate from his own projections into some sort of "mysterious" synchronicity is not only unwarranted but borders on the ridiculous.

While I can well appreciate that this license plate moment meant something to Elliot in a time of need, it doesn't at all follow that such is an example of a mysterious synchronicity. Far from it—it shows, rather, how easy it is for human beings to find meaning in letter sequences. Kids do it all the time at school, but they don't then proceed to scientifically argue that something "magical" is transpiring.

In addition, a few questions arise after reading Elliot's narrative. Were the letters ACT in capitals? Did other letters or numbers surround them? Were there other license plates that Elliot looked at during this time? Did those other plates have anything significant on them as well? If not, then are those considered to be "misses?" I can come up with a whole of questions that need to be answered, but all of them are unnecessary if the real cause of the event points directly to human intentionality. To be precise, what we are witnessing is the brain's ability to find meaning in almost anything, including three letters conjoined from our alphabet placed on a metal plate riding low on the back of a four wheel car.

Let's move on to Elliot Benjamin's second example and see if perhaps this one holds up better. Writes Elliot, "I concluded my little perfect number lesson to my son and his friends by disclosing that the third perfect number is 496, and that this was the number in my e-mail address. A few minutes later, as Jeremy was driving me and one of his friends who had very actively participated in my perfect number lesson to our destinations, I noticed that the license plate on the car in front of us said, lo and behold: "496"! I immediately pointed this out to Jeremy and his friend, and we were all quite impressed and amused by this amazingly concrete display of playful synchronicity."

Again, a few questions should be asked before we proceed. Were the numbers 496 alone by themselves on the license plate? Were there any other numbers or letters surrounding 496? If so, what were they and why were they not listed? I can readily appreciate that this event holds special significance for Elliot, but there is nothing here that rises above mere coincidence. I say this because I think each of us, perhaps daily, have such oddities arise and yet we don't conjure up a metaphysical template to explain them.

That such a number has significance to Elliot is one thing. That such significance indicates something metaphysical is quite another. Elliot could just have well seen the number 6 or 28 or even 43 etched on a license plate and (given his predisposition in finding meaning in these things) used the same as illustrative of a mysterious synchronicity. This is a parlor game of intentionality and one that I have myself played over the years with great amusement.

Knowing that intentionality and probability intertwined can produce all sorts of unexpected things, I sometimes say to

9

myself before going into the Newport Beach library bookstore (a treasure trove for a book collector) that I have to get into the "Littlewood" stream. One time, I went into the bookstore and I was looking for a nice edition of Leo Tolstoy's Anna Karenina. But alas none was found. So I said to myself, "Ah, wait and see; let's play the Littlewood game." A minute or so later, the librarian brought a new trove of books for sale to be placed on the shelves. And the first book she brought out? A leather bound copy of Tolstoy's classic.

Okay, today, after I go to my favorite Vegan restaurant, Veggie Grill at the U.C. Irvine, I walk into the Newport Beach Library bookstore, but before I walk through the doors I say myself "Littlewood! Game on!"

I have this strange habit of closing my eyes when I go to the Classics section and randomly pulling a book off the shelf to see what I can unconsciously uncover. Okay, right before I go to the bookstore I was thinking about making a little movie on that article I did on Mary Magdalene. I was playing out in my head how the scenes would work, etc., since I had just finished a film on the *Kirpal Statistic*, where I pointed out how human expectations and human neurology can produce all sorts of fantastic results during meditation but which has precious little to do with the so-called mastership of the initiating guru.

As this is going through my mind, the very first book I pulled randomly off the shelf was an old novel written in the 1950s about (yes, you guessed it right) Mary Magdalene. I love this game. Of course, we tend to only recount the times it works and neglect the times it doesn't.

Do I really think like Elliot Benjamin that these book coincidences "may serve as a reminder to us that there is indeed inherent spirituality in the universe that cannot be explained by our rational scientific technological minds or brains." Of course not. To the contrary, I think it shows how human intentionality intertwined with probabilities plays out over time. I think it is the height of hubris to think that the universe is contouring to my internal whims.

But Elliot provides us with two more examples that he thinks qualify as significant,

"Last week Dorothy (my significant other) and I went camping to celebrate her birthday, and on the way back I realized that she was almost exactly twice as old as my son Jeremy, as he turned 29 two days before Dorothy turned 58. I was thinking how interesting this was to me as I was driving to my weekly tennis game, two days after Dorothy's birthday. After I stopped off to have a quick bite to eat on the way, I noticed the license plate of the car parked next to me; it was 2958!"

At this stage, one might think that Elliot is suffering from apophenia, which "is the experience of seeing meaningful patterns or connections in random or meaningless data." As Sandra L Hubscher points out in her fine article of the same name, "This brings us to a number of deficiencies in the natural human assessment of randomness. One is that randomness, by virtue of its nature, does contain some patterns. Being pattern seekers, we focus on and over-interpret these patterns."

Once again I can well understand why Elliot can get excited by noticing numbered coincidences, but he seems to have a resistance to accepting how commonplace these occurrences can be, given the powerfulness of human "patternicity." As Michael Shermer, the first person to apparently coin the term, explains in a widely cited article "Patternicity: Finding Meaningful Patterns in Meaningless Noise" in Scientific *American* (November 25, 2008):

"Traditionally, scientists have treated patternicity as an error in cognition. A type I error, or a false positive, is believing something is real when it is not (finding a nonexistent pattern). A type II error, or a false negative, does not believe something is real when it is (not recognizing a real pattern—call it "apatternicity"). In my 2000 book *How We Believe* (Times Books), I argue that our brains are belief engines: evolved pattern-recognition machines that connect the dots and create meaning out of the patterns that we think we see in nature. Sometimes A really is connected to B; sometimes it is not. When it is, we have learned something valuable about the environment from which we can make predictions that aid in survival and reproduction. We are the ancestors of those most successful at finding patterns. This process is called association

11

learning, and it is fundamental to all animal behavior, from the humble worm C. elegans to H. sapiens. Unfortunately, we did not evolve a Baloney Detection Network in the brain to distinguish between true and false patterns. We have no error-detection governor to modulate the pattern-recognition engine. (Thus the need for science with its self-correcting mechanisms of replication and peer review.) But such erroneous cognition is not likely to remove us from the gene pool and would therefore not have been selected against by evolution."

Sometimes these patterns can be very startling, much more so in fact than what Elliot Benjamin has provided for us in his article. But just because they are surprising doesn't then mean that they are indicative of a spiritual intelligence guiding the universe. I too, like Elliot Benjamin, have witnessed several amazing coincidences--the likes of which seem to defy probabilities. But therein lays the catch. Strange coincidences do happen and they happen much more often if they are coupled with a creature evolved to be a pattern seeker. Homo sapiens are gifted enough to connect dots that are unconnected, to conjoin meanings that are disjointed, and discern deeply personal messages from the universe in the letter and number sequences on license plates while driving in a car.

A few years ago I wrote the following in my online diary on *neuralsurfer.com*, "The other night I was watching a couple of DVDs which is my habit if I get off the computer early enough. None of the movies were any good, but I did find the previews of coming attractions intriguing. One of the new movies coming out from Lion's Gate is *Peaceful Warrior, which* is based on the life story of Dan Millman, a former world champion athlete and the author of several books with millions of readers. He is very well known in New Age circles and has been so for a couple of decades. After seeing the preview I was over at a friend's house who unexpectedly asked me about Dan Millman and his books completely out of the blue (as I had not mentioned watching the preview). I mentioned that I was aware of his writings, but I didn't know much more than that. Well, today, guess what? I receive a completely unexpected and unsolicited letter from Dan Millman himself asking if I would give him my permission to use some of my writings in his forthcoming memoir. He mentioned how he had read my earlier book, *Exposing Cults: When the Skeptical Mind Confronts the Mystical* (New York: Garland Publishers, 1992). Naturally, I gave him my permission and mentioned briefly the odd coincidence of him writing just now."

Now if I, like Elliot Benjamin, wished to ad hoc devise a number scale on the unusualness of the preceding synchronicity, I am fairly confident that I could massage the numbers enough so as to make it appear that these three unrelated events (seeing a preview of Dan Millman's movie, being asked about Millman the next day, and then having Dan Millman write to me) could not simply have happened by chance.

But assigning probabilities in this after-the-event way is pseudoscientific sleight of hand, since it implies a neutrality and objectivity that cannot be ascertained. That's precisely why science often necessitates double-blind experiments and rigorous experimental protocols so that the researcher's biases (consciously or unconsciously) get minimized in the translative process.

What Elliot Benjamin should champion is not his own numbering system, since that is already tainted from the start (just as mine would be as well), but one drawn from a body of disinterested scientists. But in the cases that Elliot Benjamin provides us (and even the ones that I, myself, have just proffered), this is not likely to happen because he is confusing his own intentional meaning system (Shermer's patternicity) with something he portends is beyond the rational brain.

Ironically, the opposite thing is transpiring. It is our brain and our intentionality and our meaning seeking ways that derives messages from metallic plates. This isn't Elliot Benjamin's fault, as it is the lot of almost all human beings. Given our subjective moods and desires, we have an endemic tendency to impute meaning and patterns on events which when objectively analyzed by disinterested observers have neither.

Sandra L Hubscher provides us with a brief history to these delusional tendencies of ours in her *Skeptic Dictionary* article "Aphohenia: Definition and Analysis":

13

"August Strindberg, the early 20th century Swedish playwright, chronicles in Inferno/From an Occult Diary his descent into what would likely be diagnosed as schizophrenia in modern times:"There on the ground I found two dry twigs, broken off by the wind. They were shaped like the Greek letter for 'P' and 'y'... [I]t struck me that [they] must be an abbreviation of the name Popoffsky. Now I was sure it was he who was persecuting me, and that the Powers wanted to open my eyes to my danger." This is an eerie and extreme glimpse at the propensity of the human mind to commit what the statisticians Neyman and Pearson (1933) termed Type I error. As a statistical error, it is the acceptance of a false positive, that is, believing to see a difference or meaning when the given result is attributable to chance. Strindberg, in this example, was driven to interpret the random arrangement of sticks as non-random written letters. Although he was laboring under mental illness, the tricks of his mind were not hallucinations, but over-interpretations of his actual sensory perceptions as being more meaningful than reality warranted. Brugger (2001) puts this weakness of human cognition as a 'pervasive tendency of human beings to see order in random configurations,' which Klaus Conrad in 1958 had refined and termed as apophenia, or the 'unmotivated seeing of connections [accompanied by] a specific feeling of abnormal meaningfulness.' Modern examples of apophenia (and its subset corollary pareidolia) are so numerous and sufficiently well-known to hardly need enumerating, but are amusing enough to merit repeating: Drosnin's The Bible Code, in which arrangements of letters pulled from scripture predicted events such as 9/11 (and, heads-up, an earthquake – 'the big one' – will hit in 2010), the infamous grilled cheese sandwich virgin Mary, Led Zeppelin's Stairway to Heaven crooning 'My sweet Satan,' when played backward, the face on Mars, and, apparently, psychoanalysis."

I realize that before I conclude this essay that Elliot Benjamin most likely won't see eye to eye with my argument since he thinks that I am perhaps too "scientistic" or reductionistic in my approach. But I think David Hume's maxim is an important yardstick here. Do we really think that the universe is bending the known laws of physics and chemistry (keeping in mind that license plates are made of chemical elements embedded within the known laws of gravity and electromagnetism) simply because I have lost my job or that the ages of my lover and son show up together? I think not.

Or, to put a more positive spin to this discussion, I would strongly suggest that Elliot Benjamin and others who believe in supernatural synchronicities should up their game a bit and provide us with some truly extraordinary examples of transpersonal intersections. But even here they will have to follow Pierre Simon Laplace's pertinent admonition, "The weight of evidence for an extraordinary claim must be proportioned to its strangeness." In other words, Elliot Benjamin's purported coincidences are not strange enough to invoke the transcendental and his evidences are insufficient to convince us that anything but the commonplace and the ordinary are occurring. Of course, just as I write these last lines, I was interrupted by my wife and two children who are playing a child's card game called "War." As I was explaining to each of them what the word "apophenia" means, I jokingly looked at the covered deck of cards and said, "In honor of the goddess of apophenia, I predict that the next card will be the number five."

Guess what happened? To everyone's chagrin, the next card was the number five. I must say that right after this happened I was tempted to get into my car and start looking at license plates.

FINAL REMARKS

1. I must also apologize to Elliot Benjamin for misspelling his first name (with one l instead of two) in my previous article. If I had to invoke a defense for such an indefensible lapse on my part, I think it stems from my "Occam's Razor" approach to spelling, which translated here means "do not multiply letters beyond what is necessary." This explains, albeit partially, why I tend to prefer to spell Occam (with only five letters) versus the more common spelling of Ockham (with six letters), which in the latter case according to my reckoning goes against the very rule that he was trying to champion.

2. I do know of one fairly strange desultory decussation which may be more properly categorized as a "precognitive" synchronicity or maybe even a deceptive prophecy fulfilled. It is a story that I am naturally reluctant to tell but so strange, I believe, that it deserves retelling. Many years ago I was on sabbatical and living in London for a few months developing content for my newly founded website, the neuralsurfer. However, I was in a pickle since I had to go to France and then Switzerland for a few weeks for personal reasons. But the situation in my life at the time was such that I couldn't exactly tell the truth to those in London about where I was going and why. I had to create a cover of sorts, so I concocted a wild tale that I had to go to Greece because a cult deprogrammer from the United States had called and he wanted me to help him in providing important information about the nefarious lifestyle of a guru named Thakar Singh. I further elaborated on this tall tale and said that the organization was willing to pay me all my expenses plus a thousand dollars a day. I then said I would be back in a week or two. The story was a complete fiction. However, it served me well at the time as I needed to completely hide my real whereabouts. Several years later, when the dust had settled and I felt safer about revealing why I had to develop such an elaborate lie, I explained to those involved that I never actually did go to Greece but ended up in Paris and in Interlaken. I thought that was the end of the sordid narrative. But a few years ago, I got the most unusual and totally unexpected phone call. A well-known cult deprogrammer in the United States called and asked if I could help him with a client who was trying to extricate his father from the clutches of a dangerous cult leader.

When I asked who the guru was my jaw dropped.

"Thakar Singh," the cult deprogrammer responded.

He then asked if I could fly abroad to meet his client.

I asked somewhat sheepishly, "Where to?"

"Greece. We will only need you for a week or two," the cult deprogrammer replied.

I couldn't believe what I was hearing. But then the cult deprogrammer put the cherry on the cake when he continued,

"Dr. Lane, we will pay all your expenses and a thousand dollars a day for your expertise."

Although I ended up declining the generous offer, I was wonderstruck by the apparent fulfillment of my earlier ruse. As I told my wife the story, she too couldn't believe how my earlier fictional story could be so perfectly matched by a later, unmediated, non-fictional offer.

But even here I think we shouldn't succumb to the transcendental temptation or invoke mystical explanations. Coincidences are, contrary to what we may wish to believe, just that. Coincidences.'

3. When I was a sophomore in high school, some of my friends and I would randomly open the Bible after asking a question we needed an answer to. Usually, given our Bohemian ways, it was where to take our girlfriends on a date on Friday or Saturday nights. Inevitably, we would find passages that seemed to pertain to our concerns.

Does this mean that the Bible, like Elliot's license plates, is responding to the internal needs of young teenagers? Of course not. Rather, it is yet another example of how human beings have the ability to find meaning in almost anything. As my now deceased friend, Paul Tooher, once opined about the guru Charan Singh and his ability to find Sant Mat teachings in almost any holy book, "Dave, if you gave Charan Singh a copy of *Cosmopolitan* magazine I am 100 percent confident that he would find passages that related to the theory and practice of surat shabd yoga." Intentionality is such a powerful tool that one can extract diamonds from mud, even when there are no such diamonds and there isn't any mud.

3 | *Voodoo Voodoo and Two More Waves*

I enjoyed M.A. Rose's response to our article *Desultory Decussation*. I can well understand his skepticism of our apparent dismissal of Elliot Benjamin's license plate synchronicities. However, in a follow-up rejoinder to Benjamin's reply to our invocation of Littlewood's Law of Miracles (entitled "Apophenia and the Intentionality Fallacy"), we wrote the following,

"I appreciate Elliot Benjamin's recent attempt to justify his belief in mysterious synchronicities and his elaborations on why his personal experiences do not seem to be the result of Littlewood's Law of Miracles or even Desultory Decussation. I agree with him, in part, because a close analysis of his license plate encounters clearly points to an easier, even more rudimentary, explanation for the phenomenon. Contrary to what Elliot Benjamin may wish to argue, the details he presents shows quite clearly that the major factor determining these so-called "spiritual" intersections is his own desire to find meaning in seemingly random events. Quite frankly, Elliot is projecting and transferring his own intentions upon a series of license plates and then deriving some purpose or design as to why those letters or numbers have some special significance."

Thus, the real thrust of our critique didn't need to invoke Littlewood's Law of Miracles at all, since as we wrote at the time "A close analysis of the first four license plate synchronicities that Elliot Benjamin provides tells us more about his own predilections than it does about some transmundane occurrence. Ironically, the mathematics that he invokes to objectively substantiate the improbability of his encounters are, in fact, a complete ruse. Probabilities and the like have actually nothing to do with the synchronicities, since the real issue at hand is Elliot's own pattern seeking. What is at work here is the intentionality fallacy, where we as subjective creatures conflate our subjective needs and wants with outward events, mistakenly believing that the latter is literally contouring to our internal forms of awareness. Elliot Benjamin hasn't tapped into some integral networking of the divine that supersedes rational science. No, what he has uncovered is how easy it is to confuse one's neurology for ontology and then pass it off as beyond current scientific explanation. What is most important in Elliot's examples is what he leaves out, since it is in those middling details that we can uncover how intentionality, and not some heavenly agent, is at play."

Yet, regardless of whether M.A. Rose will see eye to eye with our critique, I think his contribution is important in keeping this desultory decussation alive. I personally feel that this issue, predicated on the theory of large numbers (and what fantastical intersections can transpire over time) is of elemental interest and should be a critical part of any future discussions on parapsychological claims.

Interestingly, the day I received M.A. Rose response I had put the final touches to an essay on how something akin to Littlewood's Law of Miracles (or its modified offspring, *Desultory Decussation*) appears to be at least contingently (if not fundamentally) connected to how consciously aware we are of overlapping and meaningful correlations that arise moment to moment. Perhaps the following article will be of some interest and hopefully keep this fruitful discussion (pro or con) alive.

The Magical Chant

It may have been Drainpipe at Zuma Beach or even T's at Santa Monica, but I have a distinct memory that my surf companions and I first discovered the magical mantra at Corral beach in Malibu. The ocean was nearly flat and I was out in the water with my childhood friends, Pat Donahue, Joe Dichiro and Rob Gilmore. I was 13 or 14 years old. We were all attempting to bodysurf but no waves were coming in, when all of a sudden someone in our group came up with a most infectious chant, apparently improvised right on the spot.

"Voodoo, Voodoo
and two more waves;
you know we need them
and we need them today."

We started chanting in unison, splashing the water, and laughing out loud. Then to our utter amazement, as if King Neptune had turned on his hearing aid, a set of waves approached from the horizon. Pat caught the first wave and the rest of us caught the second one which was slightly bigger. And then, almost as if responding to some hidden cue, the ocean went flat again.

I think it was right then and there that the Voodoo Voodoo chant became part of our surf lore and one which we used whenever we were in need of waves.

Of course, we never literally believed that the chant worked, but over the years we did find a number of remarkable synchronicities whenever it was invoked. It always seemed to produce the desired effect.

I bring all of this up now because this last summer I was in Waikiki with my family for a much needed vacation. The surf had been surprisingly good for most of our trip. However, near the end of our sojourn, the South Shore had gone flat.

Due to the sea's calmness I decided to take my youngest son, Kelly, out for a paddle. He was only five so the fact that there were essentially no waves worked to our advantage. I just thought we would have some fun looking at the sea turtles and the surrounding reef.

Kelly and I paddled out pretty far out to a spot called "Pops" (slang for "Populars"), since I thought we might find a tiny little wave to ride. But alas no such luck. After looking around at the surrounding beauty (with a breathtaking view of Diamond Head and the wall to wall high rise hotels), Kelly asked me why we couldn't surf a wave together. I explained that there was no incoming swell so most of the waves that were coming through were not breaking strong enough to ride.

Kelly was a bit sad since he really wanted to surf with me. It was right then I that told him of how when I was quite young my friends and we came up with a chant that when rightly repeated could generate waves. Naturally, I exaggerated a bit and added some spice to the legend, detailing the mythological lore with tidbits about how King Neptune and his crew make waves by blowing huge bubbles from under the sea whenever they have a birthday party. Kelly was wonderstruck and, to my chagrin, bought the story hook, line, and sinker.

Then Kelly exclaimed, "What is the chant? Let's do it now!"

I hesitated for dramatic effect wanting Kelly to think of how truly magical the mantra was and how only very few surfers in the world know of its power. Moreover, I was fearful of disappointing Kelly if the chant didn't work, which given the calm state of the water I thought was quite likely. Being so young, his patience for my strung out narrative, lasted all of about five seconds, so I finally gave in and explained precisely how one had to sing the chant. Kelly immediately understood and in unison we sang, "Voodoo, Voodoo, two more waves; you know we need them and we need them today." The emphasis, interestingly, is on the last syllable of Today (such as "Day ay ay ay").

Well, to my complete astonishment, right when we finished singing the chant, a series of unexpected waves came, triple anything that had come in for the past few hours.

Kelly and I didn't even have to paddle but two yards and we were gliding down the pristine face of a three footer. We rode the wave for nearly a hundred yards, hooting the whole time about our surprising good fortune.

As we paddled back to the lineup, the ocean went to sleep again and was as smooth as the sheets on our beds at the Royal Hawaiian. But Kelly was too stoked to give up now. He turned to me and yelled, "Let's do that chant again."

I tried to explain that it only works once or twice during a session and one doesn't want to overdo it, lest the mantra lose its efficacy. But Kelly wasn't buying any of my delay tactics. We chanted again, but this time a bit louder since I explained that King Neptune lives deep down in the ocean and sometimes is hard of hearing. I didn't think it was going to work at all, or at least I didn't think our chant would correlate with the natural ebb and flow of the ocean's wave production.

But I must admit that I was completely shocked when a set even bigger than the last one loomed on the horizon. We did a no paddle takeoff and this time we went left down the line and rode almost all the way to the beach.

Kelly was beyond excited and we ended up doing the same ritual several more times, each time catching waves that I didn't think even existed until we repeated the magical mantra.

Later that night as we were all having pizza at Il Lupino, Kelly wanted to know why the chant worked as it did. He sensed that there was some rationale behind the apparent magic. I told him and my other son Shaun about a famous mathematician from Cambridge University named G.E. Littlewood who had by his study of large numbers come upon a little known and little understood secret about the probabilities of a miracle occurring once every month. I wanted to explain my own take on this subject, which I called *Desultory Decussation*, but I knew it was a little too complicated for Kelly. So instead I said something slightly simpler, but nevertheless true. "Every once in a while things happen in nature, just by chance, that appear so wondrous and so surprisingly that we think that it must be due to some supernatural intervention. But on closer inspection, it turns out to be due to just the odds of how things work out. Just as when we play the card game Crazy 8's or War, sometimes an unusual sequence occurs, such that they look to be part and parcel of some guiding intelligence.

But if we play enough card games we soon realize that it is just the nature of the game that has a set of ascending numbers or values. Likewise the ocean is, in this analogy, similar to a vast card game where all sorts of hands can be dealt. Therefore, on occasion, the chanting surfer can be just plain lucky when his wishful mantra correlates with his or her object of desire—a set of waves.

What we tend to forget in this game of intended wishes is how many times it doesn't work. We only remember our "hits" and neglect how many misses there have been.

Most of what I said sailed over Kelly's head, and he very amusingly replied, "So, Littlewood and not Neptune, is why we chant Voodoo Voodoo."

I replied, "Not exactly, but he was the guy who is responsible for our understanding of how unusual things can naturally occur. Just as getting a royal flush is very rare when

23

playing poker, but it becomes distinctly possible, nay probable, if you play enough hands."

THE LITTLEWOOD GAME

It was right at this juncture that I realized that Littlewood's Law concerning large numbers (a miracle a month, as Freeman Dyson once shorthanded it) could actually form the basis of an intriguing game, but which is not played out on a board (though it could apply just as well there too), but in one's day to day life. If the matrices are true then the more aware we become of what I have termed desultory decussation (where two apparently random events intersect to form an X), the more often we should experience "synchronous" events—events so unusual and unexpected so as to seem like a "miracle." Perhaps the reason we don't experience more amazing desultory decussations in our lives is because we remain unaware of Littlewood's Law and thus unconsciously blind ourselves from all the fantastic possibilities and probabilities that await us.

I think the real reason the magical mantra "Voodoo Voodoo" worked on occasion is because it forced our little band of surfers to open up to the ocean's innumerable possibilities. Indeed, maybe that is the secret behind all such magical rituals: by invoking them we consciously awaken to nature's underlying and never ceasing game of roulette. As the advertisement for the California Lottery states it, "You cannot win unless you play."

Or, as Kelly (wise beyond his years) explained, "It isn't because King Neptune cannot hear us or is asleep, Papa. It is because Voodoo Voodoo wakes us up."

"Voodoo, Voodoo
and two more waves;
you know we need them
and we need them today."

First, I want to thank Elliot Benjamin for responding to our recent article, *Apophenia and the Intentionality Fallacy*. This is a fun dialogue and I, personally, have been enjoying thinking through Elliot's reasoning on why he believes license plates synchronicities necessitate deeper explanations than merely chance and coincidence. In addition, I want Elliot to know that I never meant to show him any discourtesy or disrespect by my occasional use of flavorful, if at times razor pointed, language. I have always enjoyed a heated discussion on controversial issues (as anyone who has seen my volleys on various forums from Eckankar to Radhasoami Studies will know), and in this regard I am grateful that Elliot has continued the discussion even if he fundamentally disagrees with the import of my criticism.

Now before I tackle two of his main points in his recent essay (mathematical probability and his invocation of quantum entanglement), I decided to take up Elliot's own experiment and do it myself. We are currently on school holiday so I have a bit more free time than usual. So, I said to myself, "Elliot finds this number 496 to be significant. Other mathematicians do as well (my wife, included, as she is a math whiz and once was asked to teach statistics at UCSD). Why not see if I too can discover similar number patterns just as he reported, with the most recent sighting being in the Caribbean."

As Elliot himself explains further, "But I believe that when we are studying the deepest realms of human experience, subjectivity becomes essential to truly gain understanding of what these experiences are all about. This was one of Ken Wilber's key points in his description of experiential knowledge in his book *Eye to Eye*, and it is also a foundation of what is referred to as 'extended science'."

In other words, I reflected to myself, "Do the experiment and see what happens." With this intention in mind, I focused on the number 496 and said, "Be aware. Be open." Instead of using license plates, I simply selected any random piece of paper that had number sequences on them. And to my amazement, the very first thing I picked up was a receipt from Don Diego's Mexican Restaurant in Indian Wells (they have a wickedly good potato taco) that had the numbers 496 in bold. A direct hit and in my first try? What are the odds for that? I mused. Before I could continue in this exercise, I had to retrieve my checkbook from the car in order to write a check for the pool man. As I pulled out a check, I couldn't help smiling to myself, the check number was 496. This, I thought was too much. I was in the zone, which reminded me of how a gambler gets on a lucky streak. Right then I glanced down at my driver's license and laughed to myself out loud and realized I was seeing 496 in almost anything. I was born on 04, 29, 1956. Take the last digit of each sequence and you get (yep, you guess it): 496. I felt as if Rod Serling was going to knock at my door any minute and say in his eerie smoke laden voice that I had just crossed over to the "Twilight Zone" (and I think the pun here was unintended).

But right after I did this, I realized that Elliot had focused primarily on license plates so I wasn't precisely following his protocols but something parallel to that. This got me to thinking that maybe I should just go to the computer and see precisely what numbers the DMV gives out. Perhaps there are certain common number clusters they give out which can account for recurring sequential patterns. So I typed in the letters DMV into the Google search engine and randomly selected a few websites. When I opened this one up I was awestruck yet again:

N C License Plate Agency
Call: (919) 496-4655

Now as I mentioned in the earlier article we wrote on this subject, I have had a number of very odd synchronicities in my life. And they were significantly more impressive (at least to me) than what just transpired with my little experiment.

It seems fairly obvious to me that what we witnessing here is how human patternicity (looking for a pattern or a meaning in apparently random events) intertwines with probability. As I suggested in *Apophenia*, anyone, anywhere and at anytime can play this parlor game and more often than not unearth some remarkable results.

However, Elliot Benjamin seems convinced that the probability of his seeing the number 496 is far too unlikely to be reduced to chance and intentionality. Here is the crux of his argument:

"So I am back in the Caribbean and I am walking past a few cars and the first license plate I notice says '4696.' I do some quick mathematical calculations and come up with something like a probability of perhaps 1 in 2000, utilizing the same kind of probability assumptions as I did in my Synchronicity and Mathematics article and taking into account that the "496' is not in succession. But then in a few minutes I see the '496" in succession at the top of the pile of license plates in a novelty store, and I'll assign the probability of something like 1 in 10,000, taking into account there are 6 slots of possibilities. Finally, since these two events are mathematically independent to the best of my knowledge, I multiply the probabilities together to arrive at $(1/2000) \times 1/10,000) = (1/20,000,000) = 1/20$ million. As I described in my previous *Synchronicity and Mathematics* article, this is the kind of problem I have with explaining highly unusual events completely by chance and coincidence, and why I continue to be open to alternative explanations, in whatever terms one is comfortable in using— science, spiritual, etc."

There are several problems with Elliot's probabilistic premises, not the least of which is that assigning probabilities necessitates strict parameters and controls. If Elliot wishes to have us seriously regard his self-reported stories as suggestive of something that can withstand scientific scrutiny, he has to set up strict and clear protocols on precisely what he is trying to measure. He provides us with neither. For instance, saying that one found the number 4696 and then assigning a probability factor to it (such as when he says, "1 in 2000") doesn't make any sense, since any number he saw could be given (with this type of methodology) the same odds. No, what should be done before assigning any probability to a number seen on a license plate is a clear rationale about what exactly is being tested.

So, for example, Elliot should write down on a piece of paper exactly what he is looking for on his initial drive in the Caribbean. Is he looking for 496 before he goes out for a drive? Getting a "series of three or four numbers or letters within a full license plate depiction of 6 or 7 possible slots" (such as 496) isn't as difficult as he assumes. But one must be absolutely clear beforehand about what exact sequence one is looking for. You cannot scatter shot look at license plates and then ad hoc choose differing combinations (such as ACT or 496) on the basis of personal needs or whim. One should be exceedingly precise beforehand about what three number sequence would constitute a "hit."

Ironically, by allowing for 6 or 7 placeholders for a three number sequence we actually dramatically increase (not decrease) the probability of finding such a combination. In the case of the number 496, the odds of finding that sequence on a license plate with 6 or 7 placeholders isn't improbable at all. It is to be expected, especially if we allow more cars in our survey. Now to see the number 496 on a license plate with 6 or 7 placeholders and then to see a different license plate with the number 496 on a license plate with 6 or 7 placeholders would constitute a nice coincidence. Ironically, in mathematics, this is much more common than one might suspect, particularly if the sample size is increased.

The best example of this three number match that I know of is called the *Birthday Paradox*. In my upper division Science and Religion course at CSULB and my Critical Thinking class at MSAC, I usually introduce this wonderful mind teaser to my students during the sixth or seventh week just as they are getting irritated with midterms coming up.

With some humor, I look around the class and ask, "What do you think the odds are that two students in here have the same birthday, keeping in mind that a typical year has 365 days in it?" Given that there are usually only 30 or so students in my course, most of them respond that the odds are not very high.

I then dramatically exclaim (doing a fairly awful Uri Geller imitation) that two students in the course should have the same birthday and that if I am wrong I will buy pizza and drinks for the entire class next week.

At this point, the students are excited since they feel very confident that I am going to lose the bet. Once a more boisterous student shouted out, "Come on, Lane, you are going down. 365 days, 30 students, do the math."

I then go around the room systematically and ask each student his or her birthday. I have done this game tens of times and to the deep consternation of my students I have only lost the bet once (even though I always buy pizza the next week anyways).

When the class hears that two students have exactly the same birthday (once it so happened that the first two students I called upon had the same birthday), they seem quite perplexed. How can that be and why was Lane so confident that he would be right?

Simple answer: math. The Birthday Paradox is explained quite nicely on the website *How Stuff Works*.

"This phenomenon actually has a name -- it is called the birthday paradox, and it turns out it is useful in several different areas (for example, cryptography and hashing algorithms). You can try it yourself -- the next time you are at a gathering of 20 or 30 people, ask everyone for their birth date. It is likely that two people in the group will have the same birthday. It always surprises people!

The reason this is so surprising is because we are used to comparing our particular birthdays with others. For example, if you meet someone randomly and ask him what his birthday is, the chance of the two of you having the same birthday is only 1/365 (0.27%). In other words, the probability of any two individuals having the same birthday is extremely low. Even if you ask 20 people, the probability is still low -- less than 5%. So we feel like it is very rare to meet anyone with the same birthday as our own.

When you put 20 people in a room, however, the thing that changes is the fact that each of the 20 people is now asking each of the other 19 people about their birthdays. Each individual person only has a small (less than 5%) chance of success, but each person is trying it 19 times. That increases the probability dramatically.

If you want to calculate the exact probability, one way to look at it is like this. Let's say you have a big wall calendar with all 365 days on it. You walk in and put a big X on your birthday. The next person who walks in has only a 364 possible open days available, so the probability of the two dates not colliding is 364/365. The next person has only 363 open days, so the probability of not colliding is 363/365. If you multiply the probabilities for all 20 people not colliding, then you get: 364/365 * 363/365 * ... 365-20+1/365 = Chances of no collisions. That's the probability of no collisions, so the probability of collisions is 1 minus that number."

I fully realize that Elliot Benjamin could argue that his alleged synchronicity has different odds than the Birthday Paradox. I agree. However, his assignation of probabilities is post hoc and not a priori and since he doesn't set up strict (and objective) guidelines about what precisely constitutes a hit beforehand with proper protocols in place (so someone from the outside would readily agree with his methodology), we are left with intentionality and patternicity as the key linchpins in his license plate experiments. As I pointed out in *Apophenia*, Elliot Benjamin's license plate synchronicities reveal more about him than about the strangeness of the world.

For instance, just today my wife Andrea and I were having lunch at Native Foods, our favorite Vegan restaurant, in Palm Desert, and as she was going through her email on her iPhone, I thought I would go outside and look at license plate numbers and see what interesting patterns I could find. However, before I ventured outside I said to myself, "How likely is it that I can find two cars with the same exact three number sequence?" Elliot claims that he found two license plates that had similar numbers in the Caribbean: the first was 4696 and the second one was 496 at a novelty shop. This very much impressed Elliot and he gives the odds as 1 in 20 million of this happening by chance.

How and why he arrives at these specific odds is itself odd and doesn't hold up under closer scrutiny for a host of reasons, some of which I previously listed. I am tempted to call Elliot Benjamin's method "Voodoo statistics" but before I crib and slightly alter George Bush's famous criticism of Ronald Reagan's economic strategy, I think it is best that I bite my tongue first. I respect Elliot Benjamin and, as my wife Andrea pointed out, he has a wonderful way with words, even if I may disagree at times with what they portend.

So I walk out to the parking lot ruminating on what the odds would be to find two license plates that have the same three numbers in sequence. And lo and behold as if Mr. Littlewood himself was guiding the proceedings (or was it Carl Jung calling down from the Collective Unconscious?) I find two completely different cars parked right next to each other with the exact same three numbers in sequential order: 895 and 895.

This is unbelievable, I thought. What a strange coincidence. Nobody is going to believe me. So, I pulled out my own iPhone and not only took pictures verifying what I found, but even video taped it. Thankfully I did so quickly, because just when I stopped shooting, one of the cars drove away.

Do I think what just happened (to cite Elliot's words referring to his own coincidences) is "not necessarily beyond scientific explanation if one enters the realm of quantum physics, where if my limited understanding of quantum physics is correct, thoughts can indeed affect physical realities and 'spooky action at a distance' is the norm."

No. I think it was a fun coincidence and nothing more. I don't for a second believe that we need to invoke quantum entanglement to explain what can already be fully understood by simple math and statistics. There is nothing spooky going on if what is happening can be explained by number theory intersecting with human intentionality and meaning seeking.

I think we can all find synchronicities in our lives, especially if we consciously intend to seek them out. I want to extend my thanks to Elliot Benjamin for giving me the impetus to do some of my own amateur sleuthing in license plate correlations. I genuinely wish I could side with Elliot Benjamin here, because then I could take this new found paranormal skill to Las Vegas and see if I could have the roulette wheel ball match my chosen number and thereby exponentially increase my wager. But alas such psychic skills have yet to pay out at gambling facilities.

In conclusion, I think readers should be forewarned before venturing out and trying their hands (which should remain, lest we forget, on the steering wheel) at finding synchronous license plate numbers. Once a number gets in your head (and following Dawkins' and Blackmore's memetic infection theory) it can be difficult to let it go. I feel like the main character in

31

Jorge Borges' classic short story, *The Zahir*, where he gets completely obsessed with a 20 centavo coin and can think of nothing else until he reaches the conclusion that he will either go completely mad or find God as a result.

5 | *You Are Probability*

"Mathematics takes us into the region of absolute necessity, to which not only the actual word, but every possible word, must conform."

--Bertrand Russell

As I glance around the room, I notice that after lecturing for an hour and a forty-five minutes straight on quantum theory, desultory decussation, and Wolfram's new kind of science, I see a Krispy Crème glaze descend over a few of my student's eyes. The fantastic implications of *chance and necessity* (to echo the title of biologist Jacques Monod's famous 1970's book) appears either to have gone over their heads or, more likely, seems of little practical consequence in their day to day lives.

I then try to draw out more clearly how understanding probability can radically alter how one views life. Imagine in this moment that you have a California Lottery "scratcher" ticket and as you systematically scratch off your numbers and their adjoining prizes you realize in the middle of class that you have won a mega jackpot of 5 million dollars. What would you do? I suspect that most of my students would stand up and leave the room there and then. One thing is for sure, however: it would wake them up and give them a huge and intoxicating adrenalin rush.

The winning student might later ruminate about his or her good fortune and reflect upon how lucky they were (given the astronomical odds against them—1 in 2,400,000) in securing that particular ticket.

I give this illustration to my students because a winning lottery number exponentially pales in comparison to the odds against them being alive and breathing (even if they nod off a bit here and there) at this very juncture in history. But in order to appreciate the anomaly of one's existence it is necessary to get a deeper understanding of the theory of large numbers.

The very fact that you are alive reading this essay is beyond any moneyed lottery you will ever enter.

Just think of your father's sperm as a starting off point. A usual male produces about 100 million sperm per ejaculation. Only one of those sperm will survive the arduous journey to its terminal apex. How many sperm does a male produce in, say, an 80-year life span? No precise count is possible, since it varies with each individual, but one can roughly estimate the number to be around 500 billion or perhaps more impressive sounding as a ½ trillion. If your own father had five children, this would mean that just in terms of sperm, you are a 1 in a 100 billion winner! Couple this with the rarity of your mother's egg (of the nearly half million follicles where only about 400 or so will become viable) and the very fact that you are alive reading this essay is beyond any moneyed lottery you will ever enter.

But this is only an infinitesimally small fraction of the monumental odds against you being here since one has to factor in all the preceding ancestors who came before. Roughly speaking, and depending on how many children were sired, the odds of 1 in a 100 billion doubles every generation, so that if you trace your lineage back to Africa 85,000 years ago, the odds become ever more daunting. This too, of course, is but a smidgen when one realizes that each of us trace our evolutionary past from humans to single cell organisms—a journey stretching back to the origins of life on this planet some 4.5 billion years ago. And even this is only 1/3 of the story since the atoms that comprise us have their basis in a history that is nearly 14 billions years old.

A visual infographic entitled "What are the Odds?" (which has gone viral on the Internet) puts the number against you being alive right now as innumerably greater than all the particles in the universe. This has led erstwhile sober scientists to theorize that the universe appears to be consciously designed with humans ultimately in mind. Naturally, such statistical comparisons can from the very start be misleading given that we are already alive and one can, if he or she so desires, do relatively the same odds for anything on terra firma, including the astronomical odds against that bottle of Coca Cola your uncle drank last night.

Yet, there is no getting around the fact that evolution is a universal version of the *Hunger Games* writ large. The survival of our distinctive genetic code over eons of time is a remarkable testimony to its fitness and adaptability. But what is even more remarkable is how much luck was involved in our temporarily resisting the 2nd law of thermodynamics. In other words, regardless of whether we calculate the odds for a wild grain of rice, a bottlenose dolphin, or a Tibetan monk, to live—even momentarily—on this third planet from the sun is a rarity enjoyed by a diminishing few.

In sum, you are a probability avatar. It is as if (metaphorically speaking) we are the result of some cosmic poker game where all the players are blind and where the winning hand is both selected and randomly determined. Perhaps the universe is built upon a mathematical superstructure such that all that we see around is the result of numbers and their relations fleshed out over time. Max Tegmark, currently a Professor at M.I.T. and a well-regarded cosmologist, argues that "Our reality isn't just described by mathematics—it is mathematics in a very specific sense."

Tegmark defines his idea as "The Mathematical Universe Hypothesis" which "implies that we live in a relational reality, in the sense that the properties of the world around us stem not from properties of its ultimate building blocks, but from relations among these building blocks."

Tegmark's idea is not new (even though how he formulates it certainly is), but has a long pedigree dating back through Plato and Pythagoras. One way to understand his view is to look at any smart phone today. Take, for instance, the iPhone 5S or its larger brother the iPad Air. There are several levels to how they operate, but we usually only access the surface level, more commonly known as the user interface. However, such ease of use is predicated upon large chunks of code derived from long and painstaking computer programs which most of us remain dutifully unaware unless we look a bit deeper and access directly the underlying operating system. Occasionally, as happened to me when I was doing rudimentary programming at UCSD back in my graduate school days, one can mess up a certain algorithm and some ungainly computer code will rear its head and be displayed on the screen itself. Thus the illusion of seamlessness is broken and one realizes (sometimes more often than one might wish) that under all the pretty pictures, movies, and fancy text, there are long and usually undecipherable strings of computational instructions. But there is an even more powerful subterranean level that most programmers never glimpse. This is where binary bits of electrical energy carry out their master's wishes in relational packets of off and on modes of behavior. How these electronic episodes behave, however, is geometrically prefigured by integrated circuits within ever shrinking silicon chips.

The iPhone 5s or iPad Air is magic to those of us playing on its "oleophobic coating, multi-touch, gorilla" glass. Most of us don't have a clue about what really makes these intelligent devices work. Analogously, Tegmark believes that physics reveals a projective world that is not what it seems.

We experience the world around us through the nine orifices of our anatomy. Yet, this is merely topical and akin to the screen on an Apple or Android device, which we can only touch, but where its constituent core remains hidden from view. If the virtual reality of a computer projection is ultimately based upon unseen digital electronic nodes, is it really a stretch to imagine, as Tegmark suggests, a universe which is in truth mathematical in structure and which coordinates quite literally the emerging patterns we see around us—from a tree, to an ocean wave, to a baby's smile?

As Tegmark remarks, "When you look around you, do you see any geometric patterns or shapes? . . . Try throwing a pebble, and watch the beautiful shape that nature makes for its trajectory! The trajectories of anything you throw have the same shape, called an upside-down parabola. When we observer how things move around in orbits in space, we discover another recurring shape: the ellipse. Moreover, these two shapes are related: The tip of an elongated ellipse is shaped almost exactly like a parabola. So, in fact, all of these trajectories are simply parts of ellipses . . . . [There] are many additional recurring shapes and patterns in nature, involving not only motion and gravity, but also electricity, magnetism, light, heat, chemistry, radioactivity and subatomic particles."

Is the universe the product of a mathematical skeletal schema, such that what we see around us is akin to a holographic projection that betrays its underlying geometric and numbered origin? While the scientific jury is still out on answering this particular query, it is interesting to note that we now have a sufficient series of telling analogies from our virtual lives (as expressed in our varying computational smart devices) to better understand the implications of Tegmark's M.U.H. hypothesis.

The theory that the universe is "math made flesh" is an instructive one, even if it only turns out to be part of a larger mosaic. We already know that Einstein's theory of relativity can only be properly understood within a geometric framework where gravity is geometry. Likewise, our deeper appreciation of quantum mechanics necessitates coming to grips with indeterminism and how probability plays an elemental part in how we not only measure the very small but also how we alter it by our array of instrumentations.

Bringing some of these ideas up to my students seems to have sparked them into a deeper appreciation of how precious life is and how studying the theory of large numbers can awaken a keener perspective on why numeracy is just as important as literacy, particularly when calculating the nearly impossible odds against one's very existence.

Tellingly, science seems to be confirming the ancient gnostic and Indian view that the world we see around is an illusory one in the sense that it betrays its real origination. As Tegmark personally concludes, *"If my life as a physicist has taught me anything at all, it's that Plato was right: Modern physics has made abundantly clear that the ultimate nature of reality isn't what it seems."*

"Which is more likely? That the universe was designed just for us, or that we see the universe as having been designed just for us?"

--*Michael Shermer*

Is the universe we find ourselves a product of intentional design? Are the laws of physics ultimately a complex recipe for life? Or, is the cosmos an incidental contingency? Are we the end product of blind and unconscious processes played out over time in a geometric wonderland of space and time?

Strangely, these questions bombarded my mind anew when I was watching a recent surf contest held in Oahu, Hawaii at the notoriously dangerous reef called "Pipeline."

It was the last and most important event in the ASP World Championship Tour that was going to determine the winner of the 2013 world surfing title. Mick Fanning was the favorite going in but in order to succeed he had to reach the semi-finals, since if he didn't Kelly Slater, unquestionably the greatest competitive surfer in history, would garner his 12th world championship if he won the event.

On the last day of competition, Mick Fanning had to first win his 5th round heat and it was an absolute nail biter as he was losing badly to C. J. Hobgood who seemed destined to block Fanning from advancing. But with less than 90 seconds remaining in the 35-minute heat, a flawless left-hander emerged and Fanning rode it to victory, securing a 9.50 (out of a possible ten points). While his family and friends went wild on the beach, Fanning knew that the World title was not secure unless he won his next quarterfinal heat against his Australian cohort, Yadin Nicol. Again, it looked like it was not Mick Fanning's day as Nicol had dominated the heat for nearly 33 minutes forcing Fanning into a desperate position needing a 9.57 to win his third world title. Watching the live webcast I (and probably most of the surfing world) thought Fanning's remarkable run was over. However, with less than two minutes left in the heat, a flawless wave showed up and Fanning rode it to perfection receiving a 9.70 score.

The unlikely coincidence of Fanning getting the right wave in the last two minutes in back to back heats was quite a remarkable feat. So remarkable, in fact, that Fanning later was quoted as saying,

"They [the waves] were pretty much exactly the same time at the end of each heat, so I don't know, whoever sent them, [but] thank you."

When I heard Fanning speak those words, "whoever sent them" they got me to reflect about how often we as humans think that certain events are "meant to be" or are part of a providential plan.

Given that Kelly Slater actually won the Pipeline Masters contest, it seemed to add more mystery to Fanning's two magical waves since if they didn't appear as they did the world title would have changed hands and Slater, not Fanning, would be celebrating his unprecedented victory.

Although a surf contest and its outcome may not seem to be instructive touchstones on the age-old philosophical and religious questions of whether the universe has a purpose or an intention I think a closer inspection will reveal that they are.

The strong anthropic principle argues that the universe is designed in such a way so as to bring forth intelligent life. It is as if the cosmos was expecting our arrival, since if any elemental particle was slightly different (hydrogen with an extra two protons and electrons, say) organic life as we presently know it couldn't exist. Fred Hoyle, the noted astronomer, claims "If one proceeds directly and straightforwardly in this matter, without being deflected by a fear of incurring the wrath of scientific opinion, one arrives at the conclusion that biomaterials with their amazing measure or order must be the outcome of intelligent design. No other possibility I have been able to think of..."

"The fundamental claim of intelligent design is straightforward and easily intelligible: namely, there are natural systems that cannot be adequately explained in terms of undirected natural forces and that exhibit features which in any

other circumstance we would attribute to intelligence."--
William A. Dembski

Other cosmologists, taking cues from Darwinian evolution, quantum mechanics, and probability theory, suggest that this universe is one of an inestimable number and thus we are here not by some overarching guiding intelligence but by chance. Alan Lightman captures this sentiment with the pithy title of his recent book, *The Accidental Universe*. As Laura Miller from *Salon.com* explains,

"The multiverse also offers a refutation of the concept of Intelligent Design. Like Intelligent Design, the multiverse is an idea that accounts for the fact that the universe we inhabit is finely tuned in various ways that permit the existence of life. If certain factors (such as the amount of dark energy in the universe) were a little bit greater or lesser — poof! But if there are an infinite, or nearly infinite, number of universes, some of which are nothing but a cold fog of evenly dispersed particles and others a single, tiny, infinitely dense point, then ours is merely one of a few universes configured so as to allow life. There's nothing particularly remarkable about our existence in it, because if it were otherwise, we wouldn't be around to remark on it. Thus, ours is an accidental universe, rather than the inexorable and inevitable result of set laws that can be discovered and understood by humanity."

I found these two polar positions (teleology and contingency) highlighted when I watched Fanning's two waves and the surfing world's reaction to them. On one hand, there are many who saw a certain precocity to what transpired in those last dying minutes, whereas as others saw just dumb luck. However, a deeper analysis reveals that Fanning's achievement was combinatorial.

First, the two waves that Fanning won with were not the product of some misunderstood mysterious process or divine miracles created from Neptune on a whim, but rather generated from a powerful pacific storm that whipped up 40 to 70 miles winds some two thousand miles away and set a well defined fetch to the north and west shores of the Hawaiian islands. Since such waves come in defined intervals, their appearance isn't at all surprising. Indeed, the very reason they

41

held the contest on that day (given that they had an extended waiting period) was due to surfline.com's very accurate forecasting which called for the swell to peak when it did. What was surprising was that Fanning hadn't caught a high scoring wave until the last two minutes. In other words, it was the drama of the event that gave the appearance of something magical, since there were excellent waves coming through the whole day.

It was Fanning's impeccable timing and skill that took advantage of the opportunity when it manifested itself. Nobody "sent" waves to him in some ontological sense (as if the ocean automatically adjusted itself to his karmic background). While it is certainly true that Fanning was in the right place at the right time (if he were five yards inside the peak or five yards down the beach, for instance, he would have been skunked), a probability matrix, and not divine intervention, better explains what transpired. More precisely, what at first glance appears to be a highly unusual and unexpected result turns out to be neither.

What is needed, of course, to properly understand apparent anomalous occurrences is a larger context and much more, not less, information. Analogously speaking, can this simpler explanation of the Fanning hypothesis also be applicable to our own seeming improbability, given that the odds against us being alive and reading right now appear to be astronomical?

Yes, if we accept contextual contingency (whether biological in a Darwinian sense or quantum-mechanical in a Hugh Everett many worlds sense), since the odds governing our eventual arrival on terra firma are not fundamentally different than the odds of two perfect waves arriving at just the right time for Mick Fanning to win the 2013 ASP world surfing championship. Odds are odds whether we are talking about the complexity of a rock, a cell, or a human being. The debate in this parameter, therefore, isn't over numbers as such but over how wide a field such statistics apply. In a mathematical cosmos chance, like gravity, appears to be universal in its import. On the other hand, given our naturally selected predisposition to find patterns and order in a highly entropic environment, it is to our advantage to impute meaning and purpose if they give us advantages in navigating and surviving this carnivorous and often unpredictable landscape. We are, in sum, combinatorial creatures who have learned to accept (no doubt at a searing price) our confused ancestry where our neurological toolkit seeks a greater meaning and purpose even as it too often realizes that whenever we look deep enough such meaning and purpose are our own projections and may have nothing whatsoever to do with how and why the universe operates as it does. "Whoever wins the lottery feels special, or even blessed, but they forget that millions of people have also played and the chances that someone won't win the jackpot decrease rapidly as successive weeks are considered. The chances for an individual are millions to one, but someone's going to get rich eventually." — *asktheatheist.com*

Seen in this light, perhaps we should echo Fanning's own agnostic revelations about the intricate mechanics of oceanography and say to the universe at large, "I don't know whoever or whatever sent me here but thanks!"

Why? Because the odds against Fanning catching those two wondrous waves at Pipeline are almost nothing compared to the odds against us being alive and having the remarkable ability to reflect upon our own causation.

We may find it soothing and reassuring to imagine that a Divine Being consciously orchestrated our emergence but such emotions don't by themselves prove that such is the case. Given the limits of our cranial capacities, we are unknowing creatures who far too often confuse the currency of our mental maps with transcendental ultimacies. While it is true that science has dramatically widened our vistas of how we understand the creation and its operating system, it would be the height of folly and hubris to imagine that our present-day cartography is the territory itself.

"There is a coherent plan in the universe, though I don't know what it's a plan for."
—*Fred Hoyle*

About the Authors

Andrea Diem-Lane is a Professor of Philosophy at Mt. San Antonio College. She received her Ph.D. and M.A. in Religious Studies from the University of California, Santa Barbara, where she did her doctoral studies under Professor Ninian Smart. Professor Diem received a B.A. in Psychology with an emphasis on Brain Research from the University of California, San Diego, where she did pioneering visual cortex research under the tutelage of Dr. V.S. Ramachandran. Dr. Diem is the author of several books including an interactive textbook on religion entitled *How Scholars Study the Sacred* and an interactive book on the famous Einstein-Bohr debate over the implications of quantum theory entitled *Spooky Physics*. Her most recent book is *Darwin's DNA: An Introduction to Evolutionary Philosophy*.

David Christopher Lane is a Professor of Philosophy at Mt. San Antonio College and an Adjunct Lecturer in Science and Religion at California State University, Long Beach. He received his Ph.D. in the Sociology of Knowledge from the University of California, San Diego, where he was also a recipient of a Regents Fellowship. He has taught previously at Warren College at UCSD, the University of London, and the California School of Professional Psychology. He has given invited lectures at various universities, including the London School of Economics. He is the author of a number of published books such as The *Making of a Spiritual Movement: The Untold Story of Paul Twitchell and Eckankar; The Radhasoami Tradition: A Critical History of Guru Succession; Exposing Cults: When the Skeptical Mind Confronts the Mystical;* and *The Unknowing Sage: The Life and Work of Baba Faqir Chand,* among others.

www.ingramcontent.com/pod-product-compliance
Lightning Source LLC
Chambersburg PA
CBHW071326200326
41520CB00013B/2879